Stewartsville

HOMETOWN HEROES COMMEMORATIVE BOOK

2026

Stewartsville Historical Society

ISBN: 978-1-957863-59-7

Cover Image: Sergey Korolev/pexels.com

Publisher's Note: The Greenwich Historical Society compiled this information with the assistance of service members' families. Punctuation, abbreviations, and even style can vary from entry to entry.

ACKNOWLEDGMENT

The Hometown Hero banners featured on the following pages were produced by Holiday Outdoor Décor of Allentown, Pennsylvania. The Greenwich Historical Society extends its sincere appreciation to Holiday Outdoor Décor and its representatives for their continued support of our Hometown Hero Banner Program over the past six years.

PARISIANPHOENIX.COM

angel@parisianphoenix.com

parisianphoenixpublishing.substack.com

@parisianphoenixpublishing
/parisan-phoenix-publishing
/parisianphoenixpublishing
@parisianphoenix
/parisianphoenix.bsky.social

Parisian Phoenix
PUBLISHING

This book is dedicated to Charlie Henshaw, USAF Veteran and member of the Greenwich Historical Society, whose vision, devotion, and countless hours of meticulous research brought together the service histories of eighty-eight veterans from the Stewartsville, New Jersey community who served our nation with honor. Charlie's passion for honoring the men and women who have served our country, and his commitment to preserving the history of Greenwich Township, have brought the stories of our Hometown Heroes to life so they will be remembered not as a passing record, but as a lasting tribute to our community's legacy of patriotism.

Images of the Hometown Hero banners and the stories preserved within these pages stand as enduring symbols in honor of those who served—Heroes, both living and fallen—whose service and sacrifice will be remembered and celebrated for generations to come. Charlie's hard work and unwavering determination in compiling this book stand as a testament to his love for his community, his deep respect for those who have sacrificed for our freedom, and his dedication to the Hometown Hero Program, ensuring that every Hero's story is told.

The Greenwich Historical Society proudly presents this work so that the names, faces, and banners of Stewartsville's Hometown Heroes may forever stand as symbols of gratitude, uniting past, present, and future generations in remembrance. Thank you, Charlie, for your vision and for helping to honor the true spirit of service that lives in our township.

HOMETOWN HEROES

ROBERT EMANUEL ALLSHOUSE

April 6, 1922 – January 20, 2009

- He served during WWII in the United States Army from November 30, 1942 thru December 1, 1945.
- After serving, Bob worked for Pfizer for 31 years.
- He married Florence E. "Betty" Morninghoff in 1948. They had one daughter, Diane Moore.
- He was a member of the United Valley Gun Club, a 61-year member of Mecsey-Bishop American Legion Post 456, the Stewartsville Grange and the Stewartsville Lutheran Church
- He was Bill Hayford's uncle.

RALPH CONRAD AMENDT

July1, 1922 – November 24, 2004

- Ralph served during WWII in the United States Army in the European Theater.
- He was a TEC 5 in Ordnance and served from January 22, 1942 until January 20, 1946.
- Ralph worked for the Mailing and Inserting Company in Lopatcong Township before going to Bell and Howell, where he worked as a mechanic for 37 years before retiring in 1986.
- He married Dorothea Strouse in 1946. They had one daughter — Deborah A. Smith.
- He was a 56-year member of Mecsey-Bishop American Legion Post 456, the Quarter Century Club of Bell and Howell, and the Railroad Historical Society in Phillipsburg.

JOHN L. BAKER

January 9, 1928 – November 27, 1986

- United States Navy, 1945–1954. Fireman 1st Class on the USS Buchanan DD-484 (destroyer), the USS San Juan CL-54 (cruiser), and the USS Bon Homme Richard CV-31 (aircraft carrier)
- After serving, he was a dispatcher in the Bethlehem office for the Lehigh Valley and Conrail railroads for 30 years, retiring in 1980.
- John was born in Phillipsburg.
- He married Adaline Adele "Dolly" Parenti.
- They had 2 sons, Gordon in Stewartsville and Michael, and a daughter Donna Tuma in Stewartsville.
- He has 4 grandchildren, 10 great-grandchildren and 1 great-great-grandson.

LEROY CHARLES BEERS

March 30, 1924 – December 6, 2015

- Leroy enlisted in the United States Army on May 31, 1943 and served in the infantry in the European Theater during WWII.
- He was wounded by an artillery shell in January 1945 during the Battle of the Bulge.
- He was awarded a Purple Heart.
- He was medically discharged from the service in September 1945.
- Leroy was a bookkeeper at Korde Distributing in Easton.
- His sisters were Shirley Rush, Helen Binder and Verna Willis.
- He had five nieces and one nephew.
- He was Ed Rush's brother-in-law.

THEODORE L. "TED" BILLER, JR.

August 16, 1919 – July 24, 2014

- Ted served during WWII as a TEC-5 in the United States Army. He served from December 10, 1940 until July 18, 1945.
- He was overseas from December 7, 1941 – May 30, 1944. He was at Schofield Barracks, Pearl Harbor when attacked by the Japanese. He fought in many battles, including Guadalcanal. He was awarded 2 bronze stars.
- Ted was born in Flicksville, Pa.
- Before entering the service, he was a farm hand.
- After the service, he was a self-employed appliance repairman.
- He married Helen Nemeth in 1946. They were married for 68 years. They had 2 sons, the late Theodore L. Biller III and the Reverend James L. Biller and wife Susan in Stewartstown, Pa.
- He was a 63-year member of the Stewartsville Presbyterian Church where he was a trustee and elder and taught Sunday School.
- Ted was a 61-year member of Mecsey-Bishop American Legion Post 456 in Stewartsville.

EDWARD JOHN BISHOP

April 3, 1911 – May 12, 1944

- WWII, United States Army. PFC in Battery C, 339th Field Artillery Battalion, 88th Infantry Division (The Blue Devils)
- Ed died of his wounds at age 33 on May 12, 1944, 2 years and 3 days after he enlisted.
- He was awarded the Purple Heart posthumously.
- Ed was the stepson of Walter Chatham, Sr.
- He was the brother of Walter Chatham, Jr. and Grant Bishop.
- He was the stepfather of Donald Cole.
- Mecsey-Bishop American Legion Post 456 was named in his honor. Members named him an honorary charter member of the Post.

RICHARD ALAN BOBROWICZ

November 4, 1948 –

- Vietnam War. Army, SP5 in “C” Company, 228th Assault Support Helicopter Battalion, 1st Cavalry Division. He was stationed in Phuoc, Vietnam from May 9, 1968 until May 8, 1971.
- Rich was awarded a Bronze Star, Air Medal w/4 oak leaf clusters, and the Air Crewman Badge.
- He is from West Orange, NJ
- He attended Bloomfield College
- Rich worked at Merrill Lynch Pierce Fenner & Smith
- He married Gloria Mitchell and they have 3 children, daughters Cheryl Bobrowicz-Greenburg and Lisa Bachhuber, and son Richard. They have three grandchildren — Laura, Vicky and Brooke and two great grandchildren.
- He and Gloria reside in Stewartsville
- He is a member of Mecsey-Bishop American Legion Post 456 in Stewartsville.

JOHN S. BODNAR

December 20, 1943 – December 12, 2019

- John served with distinction as a Corporal in the United States Marine Corps. He was on active duty from 1962-1966 and discharged in November 1968. He was an Electrical Technician, initially stationed in Jacksonville, N.C. prior to being deployed to Vietnam.
- After the service, John owned and operated Bodco Trading Inc., a company dealing with government surplus.
- John was born in Perth Amboy, N.J. and lived there till he met his wife Pat Riley at Woodbridge High School. He married Pat on July 12, 1969, and they settled in Sayreville, N.J.
- John and Pat moved to Stewartsville in later years and were happily married for 50 years.
- He was an avid hobbiest and enjoyed flying his model airplanes and drone.
- He was a member of the Tri-County Radio Control Club in North Brunswick, N.J.
- John was the Paymaster for the Marine Corps League, Corporal Kevin J. Reinhard Detachment #189 in Avenel, N.J.

KENNETH HOWARD BROWN

September 30, 1918 – December 18, 2009

- Ken fought in WWII as a SSGT in the Army Air Corps. He served from May 9, 1942 until January 17, 1946.
- He was a tail gunner on a B-24 in the 831st Squadron, 485th Bomb Group, 15th air Force in Italy. His plane was shot down May 29, 1944 over Austria and he was captured by the Germans. He was in a POW camp in Poland for 11 months. As Russians approached, the Germans released the POWs.
- He was awarded a Purple Heart.
- He spent 5 months in the Fort Dix hospital before leaving the service.
- He worked in the Foundry Division of Ingersoll Rand for 33 years, retiring in 1981.
- He married Mary Berger. They had a daughter, Kay Brown Gary, a son, Thomas Brown and a stepson, Joseph Mecsey, Jr.
- He was a life member of Mecsey-Bishop American Legion Post 456.

THOMAS BROWN

- Tom was a Captain in the United States Army and was a helicopter pilot during the Vietnam War.
- He is a West Point graduate, class of 1973.
- He is the son of WWII hero Ken Brown and his wife Mary Berger Brown.
- He is the brother of Kay Brown Gary. Kay and her husband Bob live in Wind Gap, Pa.
- Tom lives in Colorado.

DONALD VINCENT BURD

March 25, 1950 –

- Donald fought in the Vietnam War as a SP4 in the United States Army, Battery B, 4th Battalion, 39th Infantry Regiment, 3rd Army. He served from October 13, 1969 until September 3, 1971.
- After the service, Don became a member of the IBEW local union102, a licensed electrical contractor and owner of TDL Electric, LLC.
- Don was born and raised in Glen Gardner.
- He married Lola Webb on July 3, 1971. They have 2 daughters, Laura Hawk and Lisa Burd. They have 6 grandchildren, Payton, Gavin, Mason, Hayden, Kayla and Taylor, and 2 great-grandchildren, Juniper and Everette.
- Don and Lola volunteer with Habitat for Humanity, and they are founding members of the nonprofit Project Be A Hero.
- They are long-time residents of Asbury and live in the house designed and built by Don with help from family and friends.
- Don is a member of Mecsey-Bishop American Legion Post 456 in Stewartsville.

PETER R. "PETE" BUSCH

September 1947 –

- Pete fought in the Vietnam War as a SP 5 in the United States Army.
- He was in the 101st Airborne and served from March 1970 until May 1971.
- He was awarded the Bronze Star
- Pete is from Basking Ridge.
- He graduated from Ridge High School, where he belonged to the chess and math clubs, MIT and Stevens Institute of Technology.
- He married Beverly G. Thompson on April 15, 1972. They have been long-time residents of Stewartsville.
- Pete is a 35-year (2025) member of Mecsey-Bishop American Legion Post 456 in Stewartsville.
- He was also a member of the Greenwich Township Board of Education

MICHAEL J. "MIKE" CAFFREY

- Mike served in the United States Navy from February 2, 2020 until February 2, 2025.
- He separated from the Navy as an E-5, Petty Officer 2nd Class.
- He was on the first crew of the USS Frank E. Peterson, Jr. DDG 121 a guided missile destroyer. That crew commissioned the ship and sailed it through the Panama Canal to Pearl Harbor.
- Mike grew up in Stewartsville.
- He graduated from Phillipsburg High School in 2013.

HARRY CARHART

January 14, 1927 – December 26, 1981

- Harry served in the United States Navy from January 10, 1945 thru August 14, 1946 during the latter part of WWII.
- He was a Ships Serviceman 3rd Class. He managed all shipboard retail and service activities (store, barbershop, laundry shop, tailor shop, etc.) on the USS Glennon, DD-620, while serving in the Atlantic and Mediterranean.
- After leaving the service, Harry worked in heating and plumbing for RH Hahn & Sons in Easton before opening his own business.
- He was born in Franklin Township and raised in Stewartsville. He went to Phillipsburg High School.
- Harry married Dorothy Nenow. Their children are Nancy, Robert, Bonnie and Cindy.
- He drew the plans and helped construct the building for Mecsey-Bishop American Legion Post 456 in Stewartsville. He was one of the charter members of the Post.
- He was a volunteer fireman with the Stewartsville Volunteer Fire Company, Greenwich Township Treasurer, and a Mason with the Masonic Temple #52 in Phillipsburg.

LEO PAUL CASEY

October 17, 1927 – August 2, 1988

- Leo served during WWII in both the Merchant Marines and the U.S. Coast Guard. He was a Seaman on the SS Sea Flasher. He served from May 14,1945 thru March 2, 1946.
- While serving in the Merchant Marines, his ship transported troops and cargo from San Francisco, Portland and Seattle to ports in the Western Pacific.
- Leo was born in Phillipsburg and attended St. Philip and St. James High School on South Main Street. He played baseball and was a member of the yearbook staff. He was awarded his high school diploma early so he could enlist.
- His father, Edward Casey, served as a Seaman 2nd Class in the U.S. Navy during WWI, and his oldest brother Charles, was killed in 1944 at Ormoc, Leyte, Philippine Islands while serving in the United States Army.
- After serving in the military, Leo was employed by Bell & Howell in Phillipsburg as a quality control inspector. He worked there for over 38 years until his untimely death in 1988.
- Leo married Dorothy Ann Gornick in 1951. They have 4 children: Kathi Jinks (Roger), Patricia Miller, Eileen Wieller (David), and Leo J. Casey (Valerie). Leo and Dorothy have 5 grandchildren and 4 great grandchildren.
- Leo and Dorothy became permanent residents of Stewartsville in 1964, raising their children on New Village Road.
- Leo was a life-time member of St. Philip and St. James Church.

WALTER DEHAAS CHATHAM, JR.

February 1, 1920 – October 1, 1989

- Walt fought in both WWII and Korea in the United States Army Air Corps and then the USAF. He served from August 12, 1939 until August 12, 1953.
- He was a SSGT in the 28th Bomb Squadron, 19th Bombardment Group, V Bomber Command. He was captured when the Japanese invaded the Philippines. He was a POW for 3½ years. He survived the Bataan death march, escaped, joined and fought along side the Philippine guerrillas. He was awarded the Silver Star and a Purple Heart.
- He was a Captain in the USAF from October 13, 1953 until June 15, 1961, and served in Korea. He is buried in Arlington National Cemetery.
- He married Janet L. Henderson on Feb. 16, 1946, then Ruby Dover. His children were son James, and daughters Margaret Weatherman, Sara Basile, Patricia Collins, Betty Jean Kaufman, and Constance Ferrell. He was Frank Markle's brother-in-law.
- He was a member of the Defenders of Bataan and Corregidor, Central Florida Chapter of the American Prisoners of War, the Purple Heart Organization, the VFW Post 4287 in Orlando, and many other organizations in Florida.
- He was a charter member of Mecsey-Bishop American Legion Post 456.

DONALD WALTER "DON" COLE

March 26, 1920 – December 8, 1999

- Don fought during WWII as a Coxswain in the United States Navy. He served from May 30, 1944 until November 19, 1945. He was on the liberty ship SS Isaac Delgado in the Pacific Theater.

- He was the son of George B. Cole and Lila Rush Cole.

- He married Edith Bowers. They were married for 59 years. They had 3 children, Elizabeth J. "Bette Jean" DeFrancesco, Donna Lee Bartlett and Robert "Bobby" Cole.

- Doris Kobler was Don's sister.

- After the service, Don was a rural mail carrier in Bloomsbury. He and his wife Edie also owned Cole's Ceramic Shop in Stewartsville. He was a Greenwich Township committeeman, member of the Board of Education, a charter member of the Greenwich Emergency Squad, an exempt member of the Stewartsville Volunteer Fire Company, a member of the NJ State Exempt Firemen's Association and president of the Stewartsville Cemetery Association.

- He was a charter member, 6-time Commander, Adjutant, Finance Officer and Service Officer of Mecsey-Bishop American Legion Post 456 in Stewartsville, the Warren County American Legion Commander, Adjutant and Service Officer, American Legion Department of NJ Sergeant At Arms and Executive Committeeman, and National American Legion Vice Chair of the American Legion Foreign Relations Council.

FRANK LESLIE COLE

July 5, 1937 – June 15, 2015

- He was a SP4 in the United States Army from 1957–1959, stationed at Fort Bragg, N.C.
- After the service, he was the owner/operator of Summit Center Garage until retiring in 1993. He owned and restored "The Old Farm" on North Main Street in Stewartsville, making it into a mini park with antique farm plows displayed in the field. It was described as a "virtual historic treasure."
- He married Diana Linda Masker. They had a daughter, Laura Garguilo, son Robert Cole and stepson Christopher Tasiopoulos.

RUSSEL EDGAR "BUCKET" COLE

August 1, 1926 – June 27, 2015

- WWII Seaman 1st Class on the aircraft carrier USS Franklin CV-13. His ship was hit by Japanese kamikazes, severely damaged, and had to be towed back to Pearl Harbor.
- He was born in Still Valley, Pohatcong Township.
- He graduated from Phillipsburg High School in 1944.
- After the service, Bucket became a dairy farmer for 25 years.
- He married Margaret Paff. They had two daughters, Debra Cole and Donna Diehl.
- He was Zeke Savacool's cousin.
- He was a charter member and 68-year member of Mecsey-Bishop American Legion Post 456. Also, a member of VFW Post 6701, the Stewartsville Grange #121, the Stewartsville Area Seniors, the Museum Association USS Franklin CV13, and the Pohatcong Valley Semi-wild Gun Club.

EVAN CORMAN

- United States Army. After serving in the Army, Evan joined the Army Reserves where he was promoted to the rank of Captain.
- He is in the Special Operation Unit at Fort Dix and is currently working at Rutgers University as a Professor of Military Science.
- Evan grew up in Stewartsville, graduated from Phillipsburg High School in 2016 and Shippensburg University in 2020. He was commissioned through the ROTC program at Shippensburg.
- He earned a Masters Degree in Museum Studies from John Hopkins University in 2025.
- Evan's parents still reside in Stewartsville. He has two brothers, one in Lopatcong and one in Hamilton, N.J.
- He married Lauren Rapsis, another Stewartsville native, in 2023. They live in Easton.
- He is a member of the NC Archeological Society and volunteers at the Sigal Museum in Easton.
- He participated in the annual D-Day celebration in Normandy in 2025 by jumping with the 101st and 82nd Airborne units.

RICHARD F. "DICK" DEBOSH

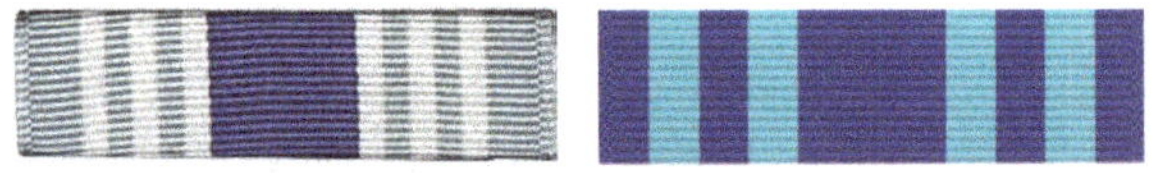

- Dick served in the United States Air Force during both the Vietnam and Cold Wars. He retired as a MSGT. He served in the 6937th Communications Group, the 6948th Electronic Security Squadron, the 6921st Security Wing, the 6989th Security Squadron and the Security Services (2 years in San Antonio). He served overseas in Thailand, Japan and Pakistan.

- He was awarded a Bronze Star, Vietnam Service Medal, Air Force Overseas Ribbon – Long, Meritorious Service Medal, Air Force Longevity Service Medal, and Good Conduct Medal.

- After the service, he worked for the US Postal Service.

- He was born in Phillipsburg, moved to Stewartsville in 1948, and after the service, moved back and still resides here.

- He graduated from Phillipsburg High School in 1959.

- He married his wife Wan, whom he met while serving in Thailand, in 1975. they have a daughter Kathy Martin in Millersville, Pa., and a son Michael.

- He likes to read, fish and collect coins and stamps.

ROY EDWARD DOWLING

October 18, 1924 – October 9, 1999

- Roy fought during WWII in the South Pacific as a Seabee in the United States Navy. He was a Machinist's Mate 3rd Class in Construction Battalion 141, D Company, Platoon 4. He served from 1943 until 1945.
- After the service, Roy graduated from Lafayette College with a degree in Mechanical Engineering. He was an Assistant Professor of Engineering at Lafayette from 1953–1960.
- Except during the war, he worked at and retired from Ingersoll Rand from 1941–1979.
- He also worked at Air Products as a research engineer.
- Roy worked the other jobs so he would be able to buy his farm. Farming was his true love. He was named the "Outstanding Young Farmer in New Jersey in 1950.
- He married Dorothy Crouse. They were married for 56 years and had 3 daughters, Mildred A. Sanders, Patricia Buzzard, and Evelyn Baxter.

THOMAS DUFFY

- Thomas served the country during the Global War on Terror as a Sergeant in the United States Marine Corps
- He served from 2002 until 2006.
- He is a retired federal police officer.
- He has been working at Lyons Veterans Hospital for the past 13 years (2025).
- He and his wife Stephanie have 4 children, twin daughters Raven and Violet and sons Fallon and Killian.
- He likes hunting and is a fan of the Yankees and Indiana Fever.
- He has two brothers, Ivan and Neil, both of whom served in the military.

ANTHONY FERRONI

- Anthony is a PFC on active duty in the United States Army's 369th Adjutant General Battalion at Fort Bragg, N.C.
- He is currently (2026) training to become an Army Ranger.
- He graduated from Warren Hills Regional High School where he wrestled and played football and lacrosse.
- He is the son of Mike and Alex Ferroni of Stewartsville.
- His brother Michael is also on active duty serving in the United States Marine Corps.

MICHAEL FERRONI

- Michael is a Corporal on active duty in the United States Marine Corps.
- He is serving in the 2nd Marine Air Wing.
- He graduated from Warren Hills Regional High School where he played football and lacrosse.
- He is the son of Mike and Alex Ferroni of Stewartsville.
- His brother Anthony is currently (2026) serving in the United States Army.

CURTIS FILLIYOR

- Curtis is a SSGT on active duty in the United States Air Force.
- He is serving in the 343 TRS Security Forces.
- He is currently (2026) stationed at Dover Air Force Base in Delaware.
- Curtis lived in Phillipsburg and Stewartsville.
- He graduated from Phillipsburg High School in 2015.

LLOYD EDGAR FOSS

October 7, 1925 – June 12, 2016

- WWII, PFC United States Army Company B, 255th Infantry Regiment "Heart of Steel," 63rd Infantry Division "Blood & Fire." Drafted Nov. 30, 1943, while a student at Lafayette College. He served for 3½ years. He was a POW, escaped from the Germans and stole their strategic plans. After Germany's surrender, he was stationed in Berlin and rode the mail train from Berlin to Frankfort. He was captured by the Russians and escaped from them. He was awarded a Bronze Star.
- After the service, he graduated from Lafayette College in 1950 with a BS In Mechanical Engineering, and Lehigh University in 1960 with a Masters in Mechanical Engineering.
- He worked for Bethlehem Steel, Ingersoll Rand, American Newspapers Publishers Association Research Institute and was VP of Newsprint Publishing for the Hearst Corporation.
- He married Carolyn A. Schroeder in 1950. Their children are Mary Lou Unangst, Caroline M. Smith, Jennifer A. Davis, Andrew T. Foss, Lloyd E. Foss Jr, Phillip A. Foss and Charles M. Foss.
- Lloyd and other veterans established the Mecsey-Bishop American Legion Post, obtained donated land from Whiz and Elizabeth Rinehart and constructed the building.
- He was a charter member, 70-year member, and Commander of Mecsey-Bishop Post 456, and once Commander of the Warren County American Legion.
- He was President of the Greenwich Township Board of Education, a 40+ year member of the Planning Board, a member of Greenwich Township Senior Citizens, the President of the Kiwanis Club in Riegelsville, and the Chairman of the Red Cross and Cancer Awareness Drives in Greenwich.

CARL HENRY FOSTER

April 15, 1920 – June 10, 2020

- Carl served during WWII as a United States Navy shipfitter.
- He was stationed at Pearl Harbor after the attack and was responsible for repairing damaged vessels.
- He graduated from Phillipsburg High School in 1937.
- After the service, he worked for Bethlehem Steel, retiring in 1983.
- He married Evelyn Strader.
- Carl was a past Greenwich Township Mayor and Committeeman, the President of the Historical Society, and the President of the Warren County League of Municipalities.
- He was also an F&AM Mason of Phillipsburg Lodge #52, a 32nd degree Mason in the Valley of Allentown Consistory, a member of the West Jersey Shrine Club, Forks of the Delaware HI-12 Club #591, the Bethlehem Steel Square Club,the Warren County Forest 314 Tall Cedars of Lebanon, Crescent Temple A.A.O.N.M.S. of Burlington, N.J. and the Delaware Chapter #50 order of Eastern Star.
- Carl was a 70-year member of Mecsey-Bishop American Legion Post 456 in Stewartsville.

WILLIAM ACKERMAN FREY

July 13, 1925 – November 1, 1992

- Served during WWII in the United States Navy. He served from July 8, 1943 until February 18, 1946.
- He was an RM3 (Radioman 3rd Class) on the USS Monticello. He served overseas from March 3, 1944–February 9, 1946.
- After the war, he was a Quality Control Foreman at the Riegel Paper Mill.
- He married Arlene Oswald. Their daughters are Sandra Stout and Melody Curzi.
- He was Dave Oswald's brother-in-law.
- Bill was a long-time volunteer fireman in Stewartsville.
- He was a 40-year member of Mecsey-Bishop American Legion Post 456.

PHILIP EDWARD GARY

October 20, 1944 – January 4, 2018

- Phil served in the United States Navy, the United States Naval Reserve, and the United States Army Reserve. While in the Army Reserve, as a Sergeant 1st Class, he served as a recruiter.
- After the service, he was a civilian contractor at Fort Drum, N.Y. for 13 years.
- He was born in Phillipsburg, lived in Stewartsville and graduated from Phillipsburg High School.
- He married Jody Meglic. Their children are Shea Boncher, Adam Gary, Cory Gary and Taylor Alexander.
- He was a past Commander of American Legion Post 61 in Watertown, N.Y.
- Phil loved the Phillies, hat collecting, softball, bowling and brightly-colored cars.
- His siblings are veteran Bob Gary, Kathryn Nuss, Mabel Cook, Jane Rush, Jean Bradbury, and Esther Schoeneberger.

WILLARD E. "BING" GARY

May 15, 1944 –

- Bing served as an E-5 in the United States Air Force's 2017 Communications Squadron.
- He served from June 1965 until June 1969.
- Bing is from Stewartsville and now lives in Egg Harbor Township, N.J.
- He graduated from Phillipsburg High School in 1963
- After the service, he was an Air Traffic Controller until retiring.
- His wife Beverly and he have daughters Stephanie and Denise and a son William.
- He is the son of veteran Willard R. and Lillian Gary. His sisters are Beverly Ferguson and Evelyn Flannigan.
- He is a 57-year (in 2025) member of Mecsey-Bishop Post 456 in Stewartsville.

WILLARD RINEHART GARY

August 19, 1919 – August 3, 2005

- Willard fought during WWII as a Corporal in the United States Army Air Corps. He served in the 8th Air Force. He enlisted December 31, 1942, and served in the European-African and Middle Eastern Theaters.
- After the war, Willard was an electrician in the maintenance department of Ingersoll Rand for 40 years, retiring in 1981.
- He was from Stewartsville.
- He married Lillian Kussler in 1942. Their children are Beverly Ferguson, Evelyn Flannigan and veteran Willard E. Gary.
- He was a member and former deacon of the Stewartsville Presbyterian Church, the Phillipsburg Lodge #52, F&AM, the Warren County Forest #14, Tall Cedars of Lebanon, the Scottish Rite, Valley of Trenton, and a member of the Greenwich Township Emergency Squad.
- He was a charter member, a 59-year member, and 5-time Commander of Mecsey-Bishop American Legion Post 456 in Stewartsville.

GREGORY GENGARO

- SSG in the United States Army's 82nd Airborne
- He was an infantryman, Radar Operator, Team and Section Chief, and Platoon Sergeant.
- Gregory served from 2016 until 2022.
- After serving, Gregory pursued a career in law enforcement.
- He is from Stewartsville.
- He graduated from Phillipsburg High School and Gwynedd Mercy University with a degree in Criminal Justice. He was a 4-year member of the soccer team.
- Gregory enjoys spending time with family and making memories.

LEONARD ALBERT GITTINS

July 26, 1927 – March 31, 1983

- Leonard was a Corporal in the United States Army Air Corps He enlisted September 28, 1945, and was discharged December 11, 1945. He re-enlisted December 12, 1945, and was discharged June 16, 1947.
- Leonard was born in Phillipsburg.
- Prior to enlisting, he worked for Ingersoll Rand
- After the service, he worked as a clerk in the shipping and receiving department at Bell and Howell for 23 years, retiring in 1982.
- He married June E. Stettler in June 1950. They had five children, Richard (in Phillipsburg), David, Barbara and Linda (all residing in Washington), and Jeff (residing in Stewartsville).
- He passed away in 1983 from cancer at the age of 56.
- He is survived by many grandchildren and great-grandchildren.

LOUIS JOSEPH GUIDI

November 18, 1919 – May 14, 2003

- WWII, SGT in the United States Army Air Corps, 730th Bomb Squadron, 452nd Bomb Group in Scotland. A munitions worker supporting the Normandy, North France, Ardennes, Rhineland, Central Europe and Air Offensive-Europe campaigns.
- Prior to the war, he worked at a cement mill. After the war, he worked at Sanco Piece and Dye in Phillipsburg and Flock Embossing.
- He was from New Village, and lived there, Phillipsburg and Stewartsville.
- He married Regina Tiberi on June14, 1941. They had 4 children, sons Bob and his wife Patricia in Middlesex, Tom and his partner Kevin in West View, Pa., and daughters Sylvia Ruby and husband Dennis in Morrisville, Pa., and the late Marilyn.
- Louis was a member of AARP and St. Philip & St. James Catholic Church.
- He was a big fan of the New York Giants, but his favorite activity was working in his garden where he grew vegetables for canning, freezing and supplying friends and family.

ROY FRANCIS “DUSTY” HACKMAN, SR.

February 2, 1933 – April 11, 2011

- Korean War; PFC United States Army
- He was a POW during the war.
- After the war, Dusty worked for Ingersoll Rand until 1987.
- He married Patricia Garo in 1957. Their children include Tammy Hill in Stewartsville, Terry L. Wolfinger, Brenda L. Mondzak and Roy F. Hackman, Jr.
- He was a life member of the Greenwich Township Emergency Squad.

PAUL D. HAGER

February 25, 1944 –

- Paul served in Vietnam as a Sergeant in the United States Army. He was in MACV (Military Assistance Command-Vietnam) and was assigned to the 2nd ARVN Division as an advisor.
- •He served from February 11, 1966 until December 7, 1967.
- After the service, Paul worked for M&M for many years.
- He lived in Phillipsburg prior to moving to Stewartsville.
- Paul married Gloria J. Fulper in June 1964. Their sons are Paul John Hager and Michael Hager.
- He is a 37-year (2026) member of Mecsey-Bishop American Legion Post 456 in Stewartsville.

HENRY ALAN HAWK

August 17, 1924 – January 15, 2005

- Alan served during WWII in the United States Navy. He enlisted on June 29, 1943.
- He served a short time as a Seaman 2nd Class on the USS Salamonie, a fleet replenishment oiler. He then served as a Ship's Cook 2nd Class on SC-1305, a submarine chaser, beginning on October 28, 1943. After sea duty, he was stationed at Pensacola Naval Air Station as a cook.
- He was on active duty from August 1943 until August 1946.
- Before enlisting, he worked for Ingersoll Rand for a year.
- After the war, Alan was a carpenter/builder and built many houses.
- He grew up in Stewartsville, and in 1942, graduated from Phillipsburg High School where he was a star football player.
- He married Helen Frances Kinkel of Stewartsville in 1946. Their children are Shirley Rush and John Alan Hawk.
- He was a charter member of Mecsey-Bishop American Legion Post 456 and was Post Commander from 1955-1956. He helped construct the Post building.

JOHN ALAN HAWK

April 26, 1946 –

- Vietnam War; WO1, CW2; UH-1 Huey helicopter pilot in 1st Platoon, 189th Assault Helicopter Company, 52nd Battalion, 1st Aviation Brigade from December 1969 until December 1970.
- After Vietnam, he joined the Pennsylvania National Guard and was promoted to the rank of CW4. For 26 years, he flew the CH47 Chinook Helicopter and flew missions in the U.S., Central America and Iceland. In 2003, he was deployed to Kuwait and Afghanistan for 1 year in support of Operation Enduring Freedom/Iraqi Freedom.
- His military service lasted 36 years during which he received numerous medals, including the Distinguished Flying Cross.
- John was born at Pensacola Naval Air Station in Florida, grew up in Stewartsville, graduated from Phillipsburg High School in 1964, earned a BA degree from Delaware Valley College in 1968, and a Masters Degree from Central Michigan University in 1980.
- He was a Warren County Health Department Inspector and Health Officer for 35 years.
- He married Heidemarie "Heidi" Friebel in December 1967. Their sons are John, Jr., Erik and Mark.
- He is a 57-year (in 2025) member, past Commander and current Finance Officer of Mecsey-Bishop American Legion Post 456 in Stewartsville.

WILLIAM W. "BILL" HAYFORD, SR.

August 18, 1943 – April 7, 2007

- Airman 2nd Class in United States Air Force from 1962 until 1966. He served in Newfoundland.
- After the service, Bill drove trucks for Smith's Transfer for 18 years, then worked for Lehigh Valley Site Contractors before retiring in 2006.
- In retirement, he drove trucks for Rodota Trucking and Wayne Willever Trucking.
- He married Bonnie L. Hummer on May 4, 1969. Their children are Beth A. Garren and William W. Hayford, Jr.
- Bill was a member of the Easton Fleas Club, the Citizens Hose Company, an exempt member of the Delaware Park Volunteer Fire Company and a Meals on Wheels volunteer.
- He was Bob Allhouse's nephew.

LUTHER BEERS HENRY

March 5, 1926 – March 5, 2005

- Luther fought during WWII in the United States Navy. He was in the Navy's AS-LTA (Aviation Support – Lighter than Air). He served in the Atlantic Theater in VPB-2 (Patrol Bomber Squadron 2). He was a gunner on a blimp and fired on German U-Boats

- After the war, he owned L.B. Henry Electrical Contractor Company in Stewartsville for 27 years. He was the plant electrician at Koh-I-Noor Company in Bloomsbury for 13 years until retiring.

- He was born in Easton, graduated from Phillipsburg High School, and lived in Stewartsville for 51 years.

- He married Lillian Richline. Their daughters are Nancy J. Carhart, Luann T. Beatty, and Wendy S. Beyda.

- He was a Greenwich Township committeeman for 9 years and the mayor for 3.

- Luther was a 52-year member of Mecsey-Bishop American Legion Post 456, the Coast Guard Auxiliary, Flotilla #10-3 in Easton, the Veterans of Foreign Wars in Easton, and the Electric League of Easton and Phillipsburg.

- He was a Boy Scout and youth group leader and Meals On Wheels volunteer.

CHARLES H. HENSHAW

March 1, 1947 –

- Captain, United States Air Force. He had a Cold War position as the Senior Missile Combat Crew Commander at site 374-9 in the 374th Strategic Missile Squadron, 308th Strategic Missile Wing, Strategic Air Command, 1969-1973.
- After serving, worked in sales for Procter & Gamble for 39 years.
- Born in Phillipsburg, grew up in Alpha, graduated from Phillipsburg High School in 1965, Rutgers University in 1969, and has been in Stewartsville for 44 years.
- He married Suzanne Hausamann on April 5, 1971. Their 3 daughters are Christine S. Baldwin (Frank), Anne E. Feeney, (Mike) and Julie L. Henshaw (Dan Marino). They have 7 grandchildren, Liliana, Josie and Eliza Baldwin, Connor and Keira Feeney, and Charlie and Harry Marino.
- He coached softball, basketball and field hockey in Stewartsville for 25 years.
- He is a member of the Greenwich Township Historical Society.
- He is the current (2026) Commander and Historian of the Mecsey-Bishop American Legion Post 456, and a Vice Commander at the Warren County American Legion.

DAVID HERSTINE

January 8, 1931 – November 29, 2017

- David was a Petty Officer in the United States Navy during the Korean War.
- He served on the USS Roanoke (CL-145), a light cruiser.
- He was born in Stewartsville, graduated from Phillipsburg High School in 1948, and Churchman's Business College.
- He was an auditor at American Can Company until retiring. He then worked for several local printing companies.
- He married Harriet Korwek. Their children are Michael, in Minnesota, Rebecca Lynn Ugander in Colorado, and stepsons Michael and Frank Korwek, and stepdaughter Ann Coris, all in East Haven, Conn. David's sister is Carol Hawk in Phillipsburg.
- He was an avid runner, participating in several marathons and enjoyed skydiving.

HARRY J. HOCKENBURY

October 29, 1926 – April 2, 2007

- He was a Staff Sergeant in the United States Army during the WWII era.
- Harry was a weapons instructor at Ft. Knox, Ky. and Camp Blanding, Fla.
- After the service, Harry was a corrosion technician at UGI Utilities in Allentown for 42 years, retiring in 1989.
- He married Dorothy K. Buss in December 1955. Their sons are Kenneth H. and Frank "Buddy" Hockenbury. Their granddaughter is Tonya Mecsey, and their grandsons are Kenneth Hockenbury, Jr. and David Hockenbury.
- He was a member of the Blue Mountain Antique Engines & Steam Association, the Forks of the Delaware Historical Arms Society, and the Safety First Fire Company in Easton.
- He loved farming and working in the family sawmill.
- He made lead soldiers and collected stamps and coins.

DONALD RANDALL HOYD, SR.

May 30, 1922 – January 2, 1984

- He served during WWII in the United States Army.
- He served from December 7, 1941 until December 31, 1946.
- He was born in Phillipsburg and lived in Stewartsville before eventually moving to Tuckhannock, Pa.
- After the service, he was employed by Riegel Paper in Milford for 38 years.
- He married Gladys Lattemann. Their children are Louisa Ann Miers in Stewartsville and Randy Hoyd in Tunkhannock, Pa.
- He was a 21-year member of Mecsey-Bishop American Legion Post 456 in Stewartsville and Triton Hose Company #1 in Tunkhannock.
- He was an outstanding baseball and softball player.
- Donald P. Miers Sr. is his son-in-law.
- He has five grandchildren (Donald Jr., Lisa, Gretchen, Hillary and Kyle) and seven great-grandchildren.

JOSEPH WILLIAM HULSHIZER, SR.

December 13, 1923 – September 18, 2006

- WWII, United States Navy, Enlisted May 28, 1943. Machinist's Mate 3rd Class on the destroyer USS Morrison (DD-560) in the Pacific Theater. He was wounded and earned a Purple Heart on May 4, 1945 when the Morrison was sunk off Okinawa by Japanese kamikazes. 152 crew members went down with the ship.
- After the war, he worked for Ingersoll Rand in the Rock Division for 43 years, retiring in 1983.
- He married Myrtle A. Fries on June 22, 1946. Their children are Anna R. Smith, Douglas R. Hulshizer, Nelson I. Hulshizer, and Joseph W. Hulshizer, Jr.
- He was a little league baseball coach, Cub Scout Master, and Assistant Scout Master of Troop #51.
- He was a member of the Greenwich Township Emergency Squad, a 29-year member of Mecsey-Bishop American Legion Post 456, the Stewartsville Volunteer Fire Company, the Mansfield Lodge #36, F&AM, Washington, and Master of the former Oxford Lodge #127.

JOSEPH COOPER HUMBERT. JR.

April 19, 1910 – June 23, 1991

- He was a Lt. Commander in the United States Naval Reserve during WWII. He was the ship's surgeon on the USS Harris (APA-2), an attack transport in the Asiatic-Pacific Campaigns. He volunteered for the Marine Raiders, was the battalion surgeon for the 3rd Marine Raider Company and participated in the Bouganville invasion. He served 2 tours of duty in the South Pacific.
- Prior to entering the service, he had graduated from Hobart College, and then, in 1933, from the University of Pennsylvania Medical School.
- He was a general practitioner in Stewartsville before and after the war, from 1939-1981.
- He married Mary C. Criswell, and they had two daughters — Emilie Humbert Mules and Rebecca Humbert Heck.
- He was a charter member of Mecsey-Bishop American Legion Post 456, the Greenwich Township Emergency Squad, and the Stewartsville Volunteer Fire Company.

CLAYTON JOHN JANISCH

July 21, 1925 – August 15, 2012

- He served during WWII In the United States Navy. He enlisted June 16, 1943. He was a Seaman 1st Class on the USS Stansbury DMS-8, a high-speed minesweeper until September 1943. He then served on the USS Biloxi CL-80, a Cleveland class cruiser in the Pacific Theater until March 1946.
- Clayton was born in and was a life-long resident of Stewartsville.
- He was a 1943 graduate of Phillipsburg High School.
- He was a truck driver who, at one time, was an owner/operator. He then worked for Steckel Concrete in Phillipsburg.
- Clayton married Mary Ann Durst Erdo.
- He was Susan Fleming's uncle.

JAMIE JOHNSON

- Jamie served the country by spending 22 years in the United States Army and Army National Guard.
- He retired as a Sergeant First Class.
- During his career, he was deployed during Desert Storm.
- Jamie's unit was also deployed to assist in the aftermath of Hurricane Katrina.

ROBERT T. "BOB" JONES

February 17, 1947 – December 20, 2021

- Vietnam Era, United States Navy on both the USS Norfolk and the USS Storms.
- Bob acquired a love of the sea, had fond memories of his Navy years, and was a proud veteran.
- After serving, Bob was employed by Osceola Graphics, Ingersoll Rand Cameron Pump Division and Alpha Press.
- He was born in Easton, graduated from Notre Dame High School in 1965, then joined the Navy.
- After his discharge, he graduated from Northampton Community College with a degree in Industrial Management.
- Bob married Jane Janosa on February 17, 1973.
- Bob and Jane loved travel, having gone on 30 cruises, and visited every continent except Antarctica.
- He was a member of Liberty Hose, Mecsey-Bishop American Legion Post 456, the Church of the Annunciation in Bloomsbury, the Holy Name Society and was on the Board of Directors for Easton PAL for many years.
- He enjoyed sports, loved football and especially the Fighting Irish of Notre Dame.
- He was a member of golf and bowling leagues for many years.

MICHAEL "KARBO" KARABINUS

August 3, 1970 –

- Mike served as an E-4 in the United States Army.
- He was in the 82nd Airborne and served from 1993 until 1996.
- Mike grew up in Stewartsville.
- He graduated from Phillipsburg High School where he played football.
- He lives in Phillipsburg.
- He's the son of Pat and the late Andy Karabinus of Stewartsville.
- Mike loves fishing and golf.

GORDON MILLEN KOBLER

April 1, 1922 – December 31, 2021

- Gordon fought during WWII as a SSGT in the United States Army. He was in the 95th Infantry Division, 3rd Army (Patton's Army) in the European Theater. He ws awarded the Bronze Star, Combat Infantry Badge and French Legion of Honor.
- Gordon was born in Newark Valley, N.Y., attended Phillipsburg High School for 3 years, and graduated from Rockaway High School in 1939.
- He was a self-employed carpenter until retiring in 1988. He build 60+ homes and helped construct the American Legion building in Stewartsville.
- He was the Greenwich Township Tax Collector from 1969 – 2013.
- He married Doris Cole in 1942. Their children are Dawn Barretto in Stewartsville and Kenneth Kobler.
- He was a lifetime member of the Stewartsville Presbyterian Church, an exempt member of the Stewartsville Volunteer Fire Company, a Mason (1951), Phillipsburg Lodge #52 F&AM, and Warren County Forest Tall Cedars of Lebanon #14.
- Gordon was the last remaining charter member of the Mecsey-Bishop American Legion Post 456 and was a member for 74 years.

FREDERICK HALSTEAD CECIL LAWTON

December 23, 1892 – January 2, 1987

- WWI Era, Cadet Pilot in Royal Air Force
- When teaching in Bethlehem Township, he traveled through New York State to Canada where he enlisted in the British Royal Air Force. The war ended prior to his joining the fight.
- He returned to the U.S. and resumed teaching in Bethlehem Township.
- He was born in Halifax, England and emigrated to the U.S. as a child.
- He married Florence A. Decher on July 6, 1921. Their children were Laurence, Arthur and Robert, and daughters Florence Dilts and Janet Wydner of Stewartsville.
- He was Elizabeth Gittins' great-grandfather and was living in Stewartsville when he passed away.

JAMES S. LEE, SR.

November 29, 1917 – April 23, 2007

- Jim served during WWII as a SGT in the United States Army. He was in the 577th Railway Transportation Battalion in the China/Burma/India Theater from August 31, 1942 until November 30, 1945.
- After the war, he worked for the Central Railroad of New Jersey and its successor, Conrail, for 40 years.
- He graduated from Phillipsburg High School in 1936, where he played football.
- He married Mary Belle Mann in 1941. Their children are Mary Xander, William R. Lee, James S. Lee, Nelson A. Lee, and the late Terry D. Lee.
- He owned the land at the site of the Morris Canal incline plane #9 West. He unearthed and restored to view the water turbine that powered the plane. That site is recognized as a National Mechanical Engineering Landmark.
- Jim became an expert about the canal, the railroad and local history. He wrote 2 books and lectured about the Morris Canal.
- He was a member of the Warren County Morris Canal Commission, the Greenwich Township Historical Society, the Warren County Cultural and Heritage Commission, the Warren County Board of Recreation and other groups.

CHESTER JOHNSON "CHET" MANN, JR.

August 22, 1919 – September 19, 2015

- Chet fought during WWII as a TSGT in the United States Army Air Corps. He was a radioman/gunner on the B-26B "Son of Satan" in the 555th Bomb Squadron, 386th Bombardment Group, 8th Air Force.
- His plane was shot down on November 18, 1944. He was captured by the Germans and was a POW until repatriated by Canadians. He was awarded a Silver Star, Air Medal and Purple Heart.
- After the war, Chet was an insurance agent for Prudential until retiring in 1974.
- He married Marguerite Thatcher who passed away in 1958, then Margaret Mondzak. Their children are Sandy Barringer, Donna Maine and Paul Mann.
- He was a member of Phillipsburg Lodge #52, F&AM and a Scout Master for Troop #51 at Grace Lutheran Church.
- He was a charter member and 69-year member of Mecsey-Bishop Post 456 in Stewartsville.

FRANK WILLIAM MARKLE

December 8, 1919 – April 21, 1978

- Frank served during WWII as a Corporal in the United States Marine Corps, D Battery, 2nd 155MM Artillery Battalion, A Battery, 7th 155 MM Gun Battalion, Fleet Marine Force Pacific (1945).
- He enlisted December 15, 1942, and was discharged December 3, 1945.
- Married Hazel Helen Chatham. Their daughter was Susan Gail Chatham.
- Frank was the brother of Navy veteran Richard Markle and Army veteran Roy Hackman, and the brother-in-law of Army Air Corps veteran Walt Chatham.
- He was Air Force veteran Amy Sandlin's grandfather.
- Frank was a charter member of the Mecsey-Bishop American Legion Post 456 in Stewartsville and was Post Commander in 1963-1964.

RICHARD LEWIS MARKLE

February 17, 1928 – December 27, 2007

- He served as a Seaman 2nd Class in the United States Navy during WWII.
- He served on the USS Whitley AKA-91 (attack cargo ship) in the European Theater.
- After the service, he was a core maker in the Foundry Division of Ingersoll Rand for 34 years, retiring in 1990.
- He married Margaret "Peg" Cutsler June 12, 1948. They had a daughter, Fran Bottos and two sons, Richard, Jr. and Frank.
- He had a sister, Doris Forder and brothers Frank, Robert and Donald Markle and Roy Hackman.
- He was a member of the Stewartsville Volunteer Fire Company, the Greenwich Township Emergency Squad, and a 10-year member of the Phillipsburg Emergency Squad
- He was a member and usher at the Stewartsville Presbyterian Church.

JOSEPH MECSEY, SR.

November 5, 1920 – May 12, 1944

- Joe fought during WWII as a PVT in the United States Army in the 350th Infantry (the Battle Mountain Regiment), 88th Infantry Division (the Blue Devils).
- Unfortunately, Joe was killed in action on May 12, 1944, when only 23 years old.
- His grave marker is in Nettuno, Italy.
- He was posthumously awarded the Purple Heart.
- Joe had emigrated from Hungary as a child and had many brothers and sisters.
- He married Mary Berger. They had a son, Joseph Mecsey, Jr.
- Grandchildren are Karen Mecsey Mulvihill and Joe Mecsey III.
- Mecsey-Bishop American Legion Post 456 was named in his honor. Members named him an honorary charter member of the Post.

JOSEPH MECSEY, JR.

August 18, 1943 – July 12, 1988

- He served as an E-4 in the United States Army's 19th Ordnance Company from 1961 until 1965.
- After the service, Joe owned and operated the Joseph Mecsey Trucking Company for a short time,
- He then became a part-time Greenwich Township police officer, before becoming the Greenwich Township Road Department Supervisor.
- He was born in and lived in Stewartsville his entire life.
- He married Ethel E. Wilson. Their children are daughter Karen and son Joseph III.
- Grandchildren are Ashley Culmone, Michael D. Mecsey and Joseph Mecsey IV.
- His mother was Mary Berger Brown, and his brother is Thomas Brown in Colorado, and sister, Kay Brown Gary.
- Joe served as Fire Chief at Stewartsville Volunteer Fire Company, was a life member of the N.J. State Fireman's Association, and an exempt fireman of the Stewartsville Fire Company.
- He was a member of the Mecsey-Bishop American Legion Post 456

DONALD P. MIERS, SR.

August 21, 1946 –

- Don served as an E-5 in the New Jersey Army National Guard from 1964–1971.
- After serving, Don was a Customer Service Tech at Embarq Telephone Company
- He retired from CenturyLink in 2007.
- He graduated from Phillipsburg High School in 1964 where he played football.
- Don married Louisa Hoyd in February 1967. Their children are Lisa Miers Hess and Donald P. Miers, Jr.
- He was a Deacon at the Stewartsville Presbyterian Church, a Mason at Mansfield Lodge #36 F&AM, a member of the Greenwich Township Historical Society and a current (2025) member of the Mecsey-Bishop American Legion Post 456 in Stewartsville.
- He has two grandchildren, Jordan and Jonathan Miers and Donald R. Hoyd Sr. is his father-in-law.

DANIEL JOHN MYERS

December 17, 1946 – January 30, 1967

- SPEC 4, Radioman, United States Army, Vietnam
- Company A, 1st Battalion, 16th Infantry Regiment, 1st Infantry Division.
- Awarded a Purple Heart for wounds suffered in hostile action near the Cambodian border.
- Danny died in a non-combat accident on January 30, 1967, 6 months after arriving in Vietnam. He was 20 years old.
- He was posthumously awarded the Soldier's Medal of Valor, which denotes acts of heroism in a non-combat situation.
- Danny was born and raised in Stewartsville. He was the son of Archie and Netti Slack Myers.
- He had a sister Netti Kathy and a brother Archie T.
- He graduated from Phillipsburg High School in 1965 and worked for J.S. Pickel Appliance in Bloomsbury prior to entering the service.
- Danny married Barbara Csordas in 1966. They had a son, Daniel Jr., who was born after Danny's death.

DAVID W. OSWALD

November 6, 1931 – September 13, 1981

- He served in the New Jersey Army National Guard from 1952 until 1981. He was the Command Sergeant Major in the 1st Battalion, 102nd Armor Division.
- He was a rifle instructor and coach of the Army Marksmanship Training Unit, and a N.J .Army National Guard recruiter.
- He was born in Stewartsville and lived in Lopatcong Township.
- He married Shirley Kliment. Their children are daughter Terri Oswald and sons David C. and Bradley Oswald.
- His sister was Arlene Frey.
- He was veteran Bill Frey's brother-in-law.
- He was a member of the Tall Cedars of Lebanon, Warren County Forest #14.

RAYMOND L. PARE

November 7, 1932 – November 26, 2010

- Ray served during the Korean War as a Petty Officer 3rd Class in the United States Navy. He served on the destroyer USS McGinty from 1950 until 1954.
- After the service, Ray was a foreman in the Rock Drill Department of Ingersoll Rand until retiring. He then worked in the Sears Service Center.
- He was born in Lowell, Mass.
- In 1949, he signed a contract to play baseball for the Brooklyn Dodgers but enlisted in the Navy instead.
- He married Claire Pacheco in 1953. Their children are Debra Loiacono in Nazareth, Tammy Kilpatrick in Wilson Borough, Ron Pare in Stewartsville, and the late Robert Pare.
- Ray was a youth baseball and wrestling coach, and an avid hunter.
- He was race car "3R's" owner and driver and raced in Harmony, Nazareth and Dorney Park Raceways.

DONALD ROSS

June 30, 1945-December 6, 2013

BERNARD ROSS

February 12, 1948-August 14, 2002

- **DON** served from 1963 until 1983. He was a Chief Boiler Technician on the USS America, the USS Guadalcanal, the USS Lindenwald, the USS Daniels and the USS John Paul Jones (DDG-32). He was on the Lindenwald with his brother Tom, and the John Paul Jones with his brother Barney.
- He moved to Stewartsville in 2000.
- He and his wife Gail had 3 sons, Kenneth Shaffer, Carl Childs and James Childs.
- He was a member and Vice Commander of Mecsey-Bishop Post 456.

- **BERNARD (BARNEY)** was an Electrician's Mate 3rd class on the USS John Paul Jones (DDG-32). At one time, he served with his brother Don on that ship.
- He was born in Harrison and graduated from Harrison High School.
- Barney never moved to this area, but both his brothers, Tom and Don, are or were members of Mecsey-Bishop Post 456 in Stewartsville.
- He and his wife Cindy had 3 children.

THOMAS J. ROSS

July 9, 1946 –

- Tom served as a Machinist's Mate 3rd Class in the United States Navy from 1964 until 1967. He served on the USS Lindenwald (LSD-6), a Dock Landing Ship.
- At one time, Tom's brother Don was with him on the Lindenwald.
- After serving, Tom worked for the Ford Motor Company until retiring.
- Tom was born in Harrison, N.J. and graduated from Harrison High School in 1964.
- He married the late Susan Roth. They moved to Stewartsville in 2003 and had two daughters, Susan Graham and Debra Rodriguez.
- Tom is a 25-year member and was a past Commander and current (2026) 1st Vice Commander of Mecsey-Bishop American Legion Post 456 in Stewartsville.

SEAN MICHAEL "HOLLYWOOD" RUANE

October 11, 1982 – January 7, 2014

- Sean was a Captain in the United States Air Force and was a Combat Rescue Helicopter Pilot.
- He rescued more than 100 people during 5 deployments to Afghanistan and Iraq.
- He held many leadership positions in the Air Force, and received many awards, including 17 Air Medals, a Meritorious Service Medal, an Air Force Combat Action Medal, and others.
- Regrettably, Sean lost his life during a night training mission while based at Lakenheath AFB in England.
- He married Rachel Roderer and they have a son, Liam.
- Sean was the son-in-law of Eveleth and Richard Roderer of Stewartsville.

ALBERT YOUNG “BERT” RUSH

January 29, 1919 – August 14, 1977

- He served in the United States Army during WWII as a Sergeant in HQ Battery 1, 510th Anti-aircraft Artillery Battalion.
- Bert was drafted July 24, 1941 and discharged January 29,1946.
- He served in the Philippines and New Guinea where he contracted malaria. Bert then received orders to go to Japan for the invasion, but after the bombs were dropped on Hiroshima and Nagasaki and Japan Surrendered, his orders were changed.
- After the service, Bert lived in an apartment over Frank Stone’s funeral home on South Main Street in Stewartsville with his wife, the late Marian Hamlen Rush. They were married on May 14, 1949.
- Bert later bought a dairy farm, starting the operation in 1951. He owned 30 Holstein milk cows, along with heifers and calves. At one time, he was honored for having the highest herd production and butterfat content in Warren County.
- He and Marian had four children, son Kevin, and daughters Kathleen Rush, Karen Rush and Kerry Smith.
- Bert graduated from Phillipsburg High School in 1938.
- He was a Charter Member of Mecsey-Bishop American Legion Post 456 in Stewartsville.
- He was Jim Sikra’s uncle.

EDWARD LEONARD RUSH

February 10, 1929 – December 7, 2012

- Ed was a SGT in the United States Marine Corps from 1948 until 1951 during the Korean War Era.
- After serving, Ed was a lineman, installer and repairman with NJ Bell Telephone until retiring.
- He graduated from Phillipsburg High School in 1948 where he was an All-State football player.
- He was inducted into the Phillipsburg Hall of Fame in 2010.
- Ed married Shirley Beers in June 1954.
- Their children include Ginny Hanisak of Stewartsville, Joni Kalapay of Forks Township, and Jen Merrick of Pohatcong Township.
- He was veteran Leroy Beers' brother-in-law.
- He was a 24-year member of Mecsey-Bishop American Legion Post 456 in Stewartsville until passing.

IRA LEEDOM RUSH, JR.

January 23, 1918 – April 2, 1989

- Leedom served during WWII as a Master Sergeant in the United States Army.
- He was drafted on March 27, 1941, the first Stewartsville resident drafted prior to WWII.
- Leedom was stationed at Pearl Harbor when it was attacked by the Japanese on December 7, 1941. While there, he worked in the war room with many generals, and taught soldiers who became affiliated with the OSS (Office of Strategic Services), the precursor of the CIA. Leedom was extremely proud of his service while at Pearl Harbor.
- After the war, he was the assistant secretary-treasurer at Victor Balata Textile Belting Company in Wilson Boro where his father was President.
- Leedom had an 8mm movie camera, carried it everywhere, and enjoyed filming many things.
- He earned his private pilot's license, owned his own plane, and enjoyed flying.
- He was a Charter Member of Mecsey-Bishop American Legion Post 456 in Stewartsville.

CHARLES JOHN RUSSO, JR.

- Charlie served in the United States Navy as a BT-2 (Boilerman Petty Officer 2nd Class) on the USS Kitty Hawk CVA-63, a US supercarrier from 1966 – 1970.
- The ship made 3 cruises to Vietnam while he was aboard.
- John was born in Phillipsburg and graduated from Phillipsburg High School in 1966, where he wrestled.
- After serving, John worked for JCP&L for 41 years as a mechanic and tool repair tech, retiring in 2011.
- John married Kathy Rae Bianco in January 1970. They have 2 daughters, Kimberly and Jennifer.
- They moved to Stewartsville in 1977 and have lived in the same house for 48 years (in 2025).
- John enjoys fishing and gardening and is involved with the Lutheran Cemetery in Stewartsville.

AMY SANDLIN

December 20, 1965 –

- During the Gulf War, Amy was a SSGT Flight Medic in the United States Air Force, 69th and 514th Aeromedical Evacuation Squadrons, from September 1984 until August 1996.
- Amy graduated from Phillipsburg High School where she was an All-State field hockey player.
- She studied sociology at Centenary University, class of 2001, and psychology at Centenary, class of 2009.
- She changed careers and studied at Drew Theological Seminary from 2012-2015.
- She became a Presbyterian minister, serving both the Delaware and First Oxford Presbyterian churches.
- Amy is now a hospice chaplain.
- She married Dean Sandlin on November 18, 2001. Their children are twins Elizabeth and Jacob, both now in college.
- Amy started the Greenwich Township field hockey program.
- She is a 34-year member of Mecsey-Bishop Legion Post 456 and a Life Member of the Stewartsville Emergency Squad.
- Frank and Hazel Markle are Amy's grandfather and grandmother, and Susan Markle is her mother.

ROBERT LUNDY "ZEKE" SAVACOOL, JR.

February 16, 1924 – December 31, 2015

- He served during WWII as a Coxswain in the United States Naval Reserve.
- He was on a Navy Armed Guard Crew on the SS Henry S. Lane from Liverpool, England to the United States from 1944 until 1948.
- After the war, Zeke was a timekeeper at Taylor Wharton in Easton for 30 years.
- He was born in Phillipsburg and graduated from Phillipsburg High School in 1942.
- He married Madeline V. Smull. Their children were daughter Gloria S. Lewis and son Ward J. Savacool.
- Their grandchildren are Dana Lewis Nyce and Robbie and Amy Savacool, and great-grandchildren Mason Robert Nyce and Colin Michael Nyce.
- Zeke was veteran "Bucket" Cole's cousin.

PETER SCHMIDT

- Peter is an E-5 on active duty in the United States Coast Guard.
- He is an Aviation Maintenance Technician/Flight Mechanic on a 65-Dolphin helicopter.
- He is currently (2026) stationed at Atlantic City Air Station.
- Peter graduated from Phillipsburg High School in 2015 and Southern Vermont College.
- He was born and raised in Stewartsville.
- He is the son of Nancy and Pete Schmidt of Stewartsville.

JEROME B. "JERRY" SEGAL

July 3, 1942 – August 2, 2011

- Jerry served during the Vietnam Era in the United States Navy from 1961 until 1967.
- He was an Aviation Electronics Technician 3rd Class on USS Hancock CV-19.
- After the service, Jerry was an auditor with AT&T in Bedminster and New York City before retiring.
- He was born in Trenton and lived in Greenwich Township.
- He married Tanya on June 23, 1989. Children are Jennifer Segal, Alexis Barsamian, David Pickel, Matthew Pickel and Kristen Todd.
- They have 9 grandchildren, Luke, Lily, Norah, Bridget, Julia, Peyton, Logan, Wyatt and Stella.
- Jerry was an avid Phillies and Eagles fan, reader, collector of baseball memorabilia and was a numismatist (study of currency-related items).
- He loved God and country.

JAMES CHARLES "JIM" SIKRA

December 29, 1943 –

- Jim retired as a Colonel in the United States Air Force, serving from 1969–1992.
- While in the USAF, Jim was involved in the development and flight testing of USAF missiles, aircraft electronic warfare systems and space re-entry systems. He served a tour of duty in technical intelligence and taught engineering mechanics at the Air Force Academy.
- His awards include the Legion of Merit, the Meritorious Service Medal with 2 oak leaf clusters, the Air Force Commendation Medal with an oak leaf cluster, the Air Force Achievement Medal and the National Defense Service Medal with Bronze Service Star.
- Jim was born and raised in Stewartsville, graduated from Phillipsburg High School in 1961, Rutgers University in 1965 with a BS in Materials Engineering, and Rutgers again in 1973 with a PhD in Materials Engineering.
- Jim married Barbara Gersitz. They have a daughter, Amy Knowlton and a son, James W. Sikra, a retired USAF Lt. Colonel.
- He is a member of the Greenwich Township Historical Commission/Society, and the John Dolak American Legion Post 446.

CHARLES J. "CHUCK" SLACK, SR.

February 5, 1935 – April 22, 2017

- Chuck served during the Korean War as a Corporal in the United States Marine Corps.
- He was in the 1st Battalion, 1st Marine Division, 5th Marines Fleet Marine Force.
- After the service, Chuck became a truck driver for most of his life, working for Branch Trucking, F&M Schaefer Brewing Company, Exide Battery Company, Keystone Snacks and Ingersoll Rand.
- He married Joan P. Unangst on August 20, 1955. Their children are Joann L. Slack Hassan and the late Charles J. "Chuckie" Slack, Jr.
- He was a member of the Mack Truck Million Mile Club, a lifetime member of the Harley Owners Group, the American Motorcyclist Association, a life member and past president of the Bushkill Valley Motorcycle Club, and the Masonic Lodge 152 in Easton.
- He was a 58-year member of Mecsey-Bishop Legion American Post 456.

TERRY M. STECKER

March 10, 1947 –

VINCIMUS SPATIUM

- Terry was a 1st LT. in the United States Army during the Vietnam War.
- He served in the 57 Signal Battalion, III Corps from June 21, 1971 until 1973.
- After the service, Terry began his banking career as a corporate lender. He eventually became Senior Relationship Manager and Vice President of First Valley Bank in Pennsylvania.
- He was born in Pennsylvania and raised in Stewartsville.
- Terry graduated from Phillipsburg High School in 1965 where he played football.
- He graduated from Norwich University in 1970 with a degree in Business Administration. He was a 3-year starter on the football team, co-captain his senior year and the winner of many awards. He appeared in *Who's Who of American College Athletes*. He was inducted into the Norwich University Athletic Hall of Fame in 1995.
- He married Susan E. Gratzel in 1971. Their children are Susannah Stecker Pilato, and sons Stewart and Grant Stecker.
- Terry is a 51-year member of Mecsey-Bishop American Legion Post 456 in Stewartsville.

HUGH ANDERSON "PETE" STONE, JR.

August 27, 1931 – April 6, 2010

- Pete was a United States Army Corporal in the 501st Anti-aircraft Artillery Battalion, Camp Hanford, Wash. He served from September 16, 1949 until December 17, 1952.
- After the service, Pete was a steel fabricator for F.L. Smidth & Company in Lebanon for 14 years, then an engineer for Ecolaire Corporation in Easton for 11 years, retiring in 1996.
- He graduated from Phillipsburg High School in 1949 where he played baseball and football.
- He married Alice McQuaid June 5, 1954. They were married for 55 years. They had two daughters, Rhonda A. Stone, at home, and Cynthia k. Durholz of Harmony.
- His brother was Robert S. "Dale" Stone, Sr. of Stewartsville and sister Joyce E. Buckley of Baltimore, Md.
- Pete was a member of the Bloomsbury United Methodist Church, a 58-year member of Mecsey-Bishop American Legion Post 456, an Associate member of Bloomsbury Hose Co. #1, the United Steelworkers of America Local 8501 and the Bloomsbury Rod and Gun Club.
- He was an avid gardener and enjoyed bowling at Warren Lanes in Phillipsburg.

ROBERT DALE "BOB" STONE, JR.

December 23, 1955 – March 4, 2014

- Vietnam Era. Bob was a United States Navy Mess Management Specialist (ship's cook) on the USS Hermitage from 1975 until 1977.

- After the service, Bob worked as a chef at several area restaurants, including Brunner's, Jonathon's Pumpernickel Pub, Harmony Manor and Station Break Restaurant. His last job was at Roxbury High School.

- Bob also ran Fat Boy Road Promotions.

- He married Brenda Hoffman March 11, 1978. Their children are son Jeffrey and daughters Melissa and Sarah. He was Ann Stone's son and the brother of Debbie Riley, Kim Stone, Denise DeMasi and Dawyne Stone.

- Bob was the President of the Stewartsville Cemetery Association, a Fire Policeman with the Stewartsville Volunteer Fire Department and a member of Gem Vac Veterans Group.

- He was a music enthusiast.

HUGH JOHN SULLIVAN, JR.

May 14, 1942 – June 5, 1965

- Hugh was a SGT in the United States Marine Corps during the Vietnam War.
- He was a Machine gun squad leader, C Company, 1st Battalion, 3rd Marine Division.
- Unfortunately, Hugh was killed in action on June 5, 1965 from small arms fire in Vietnam's Quang Nam Province. He perished while manning a machine gun trying to protect the evacuation of wounded Marines. He was 23 years old.
- He was posthumously awarded the Silver Star, Purple Heart, the Vietnam Gallantry Cross and the Combat Action Ribbon.
- Hugh attended Phillipsburg High School during his freshman year, then transferred to William Allen High School. After completing 11th grade in 1959 at age 17, he chose to enlist in the Marine Corps where he spent the last 6 years of his life.
- He married Sandra Corbitt of Charleston, S.C. Their daughter Christine Lynn was born while he was in Vietnam and was 3-months old when he perished. Hugh never met his daughter.

CHRISTOPHER “TAZ” TASIOPOULOS

- US Army, Sergeant,Military Police Investigator, Special Reaction Team Leader
- He was on active duty from 1996-2001 with the 8th Army, 142nd MP Company in Seoul, South Korea and the US Armor Center MP Company at Ft. Knox, Ky.
- In the NJ National Guard, Chris was a sergeant and Squad Leader in Detachment 2, 42nd MP Company and was activated after 9/11 to conduct security at Ft. Dix, N.J.
- Chris was born in Newton, N.J. and grew up in Phillipsburg. He graduated from Phillipsburg High School in 1993.
- He has a BS degree from DeVry University in Computer Science-Information Security.
- Chris is married, and he and his wife Samantha have four children, daughters Phoebe and Morgan and sons Orion and Quintus.
- Chris has worked for the Belvidere Police Department and the Warren County Sheriff's Office
- He is currently a Sergeant with the Greenwich Township Police Department where he had also served as a detective and a K-9 handler.
- He coached baseball for Greenwich and Pohatcong and basketball for Greenwich and Alpha.
- He is Diana Cole's son and veteran Frank Cole's stepson.

BRIAN THOMPSON

- Brian was a Corporal in the United States Marine Corps from 1995 until 1999.
- He was stationed at Okinawa in Marine Wing Support Squadron 172.
- Brian was also stationed at Marine Corps Air Station Cherry Point, N.C. in Marine wing Support Squadron 271.
- He grew up in Sparta and White Township and now lives in Stewartsville.
- He graduated from Belvidere High School in 1994.
- Brian has three sons, Trevor, Blake and Hunter.
- He works with his father at Warren Thompson and Son Roofing, Siding and Gutters.
- Brian coached his sons in lacrosse and soccer.
- He enjoys the beach and off-roading.

EDWARD GEORGE WACKERMANN

January 14, 1907– July 28, 1970

- Ed served during WWII as a SGT in the United States Army.
- He was a motor pool mechanic.
- He enlisted November 25, 1942, and was discharged August 13, 1943.
- After leaving the service, Ed worked as a mechanic.
- He was born in Brooklyn, moved to Lopatcong Township and then Stewartsville in 1929.
- He married Helen Gercie in 1945. They had 4 sons, Carl, Ed, Bill and Jim. Jim lives in Stewartsville.

EDWARD W. WACKERMANN, SR.

February 19, 1947 – September 2, 2025

- Ed was a PFC in the United States Army during the Vietnam Era.
- He was the son of Edward and Helen Vayda Wackermann.
- Ed graduated from Phillipsburg High School in 1965.
- After the service, he was an iron worker.
- He was married three times: to Renee Powell, Georgia Crouse and Debbie Wheeler. He had a son, Edward Wackermann, Jr. and stepchildren, Felicia Tomino, Georgia Weber, Albert Dilts and Wendy Dilts.
- He most recently lived in Saylorsburg, Pa.
- His father, Edward G. Wackermann, was a WWII Army veteran.
- His brothers are Jim, who lives in Stewartsville, George "Bill" Wackermann and Arnold Wheeler.
- Ed was an avid hunter and fisherman.

ROLAND FIRTH WALDBIESSER

March 30, 1895 – May 4, 1982

- Roland was a PFC in the United States Army and fought in WWI.
- He was a cook and 1st Class Gunner in the 133rd Company C.A.C. (Coast Artillery Company), January 24, 1915 – February 20, 1918, Battery A, 56th Regiment, Headquarters Detachment February 20, 1918 – March 12, 1919, 4th Company C.A.C. March 12, 1919 – June 4, 1920. He was in France March 28, 1918 until March 12, 1919. He was discharged June 4, 1920. His major engagements were the Marne offensive, Ainse Marne and Meuse Argonne. He shot down a German plane.
- After serving, he worked as a butcher.
- Roland was born in Phillipsburg and lived there, Bethlehem, Pa. and Stewartsville.
- Stepdaughter Jane Phillips and her husband Bob live in Stewartsville,

ROSS WOODROW WILSON

January 5, 1917 – May 13, 1956

- During WWII, Ross was a TEC5 in the 'United States Army.
- He was a Brakeman in Company C, 740th Railway Operating Battalion, 1st Army.
- Ross served in the Ardennes, Central Europe, Northern France, and Rhineland Campaigns.
- Before entering the service, he worked as a farmhand on a farm rented by his father, then worked for Henderson Supplee in Washington.
- He was born in Bedford, Pa., then lived in Oxford and Washington, N.J.
- After he left the service, Ross worked in the textile industry.
- Ross married Emma Elizabeth Yaple in 1939. They had a daughter, Ethel E. Cole.
- His grandchildren are Joe Mecsey III, Karen Mecsey Mulvihill, Jodie Kunkel and Bobby Kunkel.

DOUGLAS J. WREDE

- Doug is a LT on active duty in the United States Navy. Currently (2026), he is stationed in Monterey, Calif., attending Navy Postgraduate School working on his Masters in Mechanical Engineering.
- Doug recently finished an 8-month deployment on the attack submarine USS Alexandria.
- Doug moved to Stewartsville when he was 1-year-old.
- He graduated from Phillipsburg High School in 2015, where he ran cross country and played lacrosse.
- He graduated from Stevens Institute of Technology in 2020 with a degree in Mechanical Engineering.
- Doug is a member of St. James Lutheran Church
- He loves to snowboard and surf and is currently training for an ironman competition.
- Doug's mother in Barbie Gulick.

MICHAEL C. ZALADONIS

- Mike was in the United States Army Reserves from 1999 until 2007.
- He is the founder and developer of Imago AI LLC, specializing in creating innovative medical education applications for ultrasound expertise.
- He has a decade of experience in software development.
- Mike's background includes being a registered Vascular Technologist, an adjunct teacher at Misericordia University and US Army HR.
- He has a BA from Marywood University, an MBA from the University of Washington-Eau Claire and a data analytics certificate from Rutgers University.
- Mike and his wife Elizabeth Gittins live in Stewartsville.

www.ingramcontent.com/pod-product-compliance
Ingram Content Group UK Ltd.
Pitfield, Milton Keynes, MK11 3LW, UK
UKHW060120010826
14089UKWH00005B/76

9 781957 86359